HEPATITIS

THINGS YOU SHOULD KNOW
(QUESTIONS AND ANSWERS)

By Rumi Michael Leigh

Introduction

I would like to thank and congratulate you for purchasing this book, " *Hepatitis, things you should know (questions and answers)*" series.

This book will help you understand, revise and have a good general knowledge and keywords of Hepatitis and how it affects the lives of people who suffer from this disease.

Thanks again for purchasing this book, I hope you enjoy it!

Chapter 1

1) What does "hepa" signify ?

- Hepa signifies liver.

2) What does "itis" signify ?

- Itis signifies inflammation.

3) Define hepatitis.

- Hepatitis is the inflammation of the liver due to an infection.

4) What kind of infection is hepatitis ?

- Hepatitis is a viral infection.

5) Could hepatitis be fatal ?

- Yes, hepatitis could be fatal.

6) What is the largest organ in the human body ?

- The liver is the largest organ in the human body.

7) What are the functions of the liver ?

- The liver metabolizes fats, carbohydrates, protein, etc. It activates enzymes, it secretes bile, it produces albumin, it excretes bilirubin, drugs, hormones and also takes part in blood clotting.

8) Where is bile stored ?

- Bile is stored in the gallbladder.

9) Where is bile released into ?

- Bile is released into the small intestine.

10) What is the function of bile in the small intestine ?

- Bile helps to digest fat in the small intestine.

Chapter 2

1) What are the functional units of the liver ?

- The functional units of the liver are the hepatic lobules.

2) What is the function of the hepatic lobules ?

- The hepatic lobules filter blood.

3) What are hepatocytes ?

- Hepatocytes are the cells of the liver.

4) How many lobes does the liver have ?

- The liver has two lobes.

5) How many segments does the liver have ?

- The liver has 8 segments.

6) What does the liver store ?

- The liver stores vitamins such as A, D, E, K, and C and also glycogen, Iron, etc.

7) From how many sources does the liver receive its blood supply ?

- The liver receives its blood supply from two sources.

8) What are the sources of blood supply of the liver?

- The sources of blood supply of the liver are the hepatic portal vein and the hepatic artery.

9) What are the characteristics of the hepatic portal vein ?

- The hepatic portal vein is poor in oxygen but rich in nutrients.

10) What is the function of the hepatic artery ?

- The hepatic artery supplies blood rich in oxygen to the liver.

11) What is the function of the portal vein ?

- The portal vein transports blood from the intestines to the liver.

Chapter 3

1) What is the incubation period ?

- The incubation period is the period necessary for the virus to produce symptoms after the infection.

2) What type of hepatitis can cause hepatocellular carcinoma apart from hepatitis B ?

- Hepatocellular carcinoma can also be caused by hepatitis C.

3) What is cirrhosis ?

- Cirrhosis is the scarring of the liver.

4) What do scar tissues do to the liver ?

- Scar tissues reduce the blood flow to the liver.

5) What is the abbreviation RNA ?

- Ribonucleic acid.

6) What is an RNA virus ?

- An RNA virus is a virus that stores its genetic material in the form of RNA.

7) What is the abbreviation DNA ?

- Deoxyribonucleic acid.

8) What is a DNA virus ?

- A DNA virus is a virus that has DNA as its genetic material and the virus replicates using a DNA-dependent DNA polymerase.

9) What is viremia ?

- Viremia is the presence of virus in the blood.

10) Apart from a viral infection, what are other causes of hepatitis ?

- Apart from a viral infection, hepatitis could also be caused by excessive alcohol consumption, steatohepatitis, certain medications, infectious diseases, and autoimmune diseases.

Chapter 4

1) What are the main types of viral hepatitis ?

- The main types of viral hepatitis are type A, B, C, D, E.

2) What do the letters A, B, C, D, E signify in hepatitis ?

- The letters A, B, C, D, E signify hepatitis viruses.

3) How is hepatitis A contracted ?

- Hepatitis A is contracted from the fecal matter of an infected person, unprotected sex, and contaminated food or water.

4) Can the body's immune system fight effectively against hepatitis A ?

- Yes, the body's immune system can fight effectively against hepatitis A.

5) What is the incubation period of hepatitis A ?

- The incubation period of hepatitis A is 30 days.

6) Can a person get vaccinated against hepatitis A ?

- Yes, a person can get vaccinated against hepatitis A.

7) Will a person who has had hepatitis A be immune to it ?

- Normally, yes, a person who has had hepatitis A will be immune to it for life.

8) What is the percentage of chronicity of hepatitis A?

- Hepatitis A has no percentage of chronicity.

9) What kind of places are hepatitis A and hepatitis E usually found ?

- Hepatitis A and hepatitis E are usually found in places and regions with inadequate or poor sanitation systems.

Chapter 5

1) How is hepatitis B contracted ?

- Hepatitis B is contracted by body fluids, the blood of an infected person, sharing needles, toothbrush, and having unprotected sex.

2) Can hepatitis B be transferred from an infected mother to her baby ?

- Yes, hepatitis B can be transferred from an infected mother to her baby.

3) Can the body's immune system fight effectively against hepatitis B ?

- Yes, the body's immune system can fight effectively against hepatitis B.

4) Can the body's immune system fight effectively against chronic hepatitis B ?

- It depends, sometimes an antiviral medication may be necessary or a liver transplant.

5) What is the incubation period of hepatitis B ?

- The incubation period of hepatitis B is 90 days.

6) What type of hepatitis has the lowest percentage of chronicity ?

- Hepatitis B has the lowest percentage of chronicity.

7) Can a person get vaccinated against hepatitis B ?

- Yes, a person can get vaccinated against hepatitis B.

8) What are the complications of chronic hepatitis B?

- Chronic hepatitis B has complications such as hepatocellular carcinoma and cirrhosis.

9) Name a common antiviral treatment for hepatitis B.

- A common antiviral treatment for hepatitis B is interferon.

Chapter 6

1) What are the vaccines for hepatitis C ?

- There are no vaccines at the moment for hepatitis C.

2) What is the incubation period of hepatitis C ?

- The incubation period of hepatitis C is 40 days.

3) What type of hepatitis has the highest percentage of chronicity ?

- Hepatitis C has the highest percentage of chronicity.

4) Does a patient with hepatitis C always have jaundice ?

- No, a patient with hepatitis C does not always have jaundice.

5) Is a person immunized after a treatment for hepatitis C ?

- No, a person is not immunized after a treatment for hepatitis C.

6) Is it useful to treat hepatitis C if a patient is HIV positive ?

- Yes, it is useful to treat hepatitis C even if a patient is HIV positive because HIV accelerates hepatitis C to hepatitis cirrhosis.

7) Can hepatitis C heal ?

- Yes, hepatitis C can heal.

Chapter 7

1) What does D represent in hepatitis D ?

- D represent the delta virus.

2) In what condition can hepatitis D be transmitted ?

- For hepatitis D to be transmitted, hepatitis B must already be present.

3) Can a person get vaccinated against hepatitis D?

- Yes, a person can get vaccinated against hepatitis D.

Chapter 8

1) What is the incubation period of hepatitis E ?

- The incubation period of hepatitis E is 50 days.

2) What is the similarity between hepatitis E and hepatitis A ?

- Hepatitis E and hepatitis A are transmitted in the same way.

Chapter 9

1) What are the three phases and clinical features of hepatitis ?

- The three phases and clinical features of hepatitis are :

- the prodromal phase,
- the icteric phase, and
- the convalescent phase.

2) What is another name for the prodromal phase ?

- The prodromal phase can also be called the preicteric phase.

3) What are the symptoms of the prodromal phase ?

- The symptoms of the prodromal phase are :

- headache,
- skin rashes,
- nausea,
- fever,
- fatigue,
- muscle pains,
- etc.

4) What are the symptoms of the icteric phase ?

- The symptoms of the icteric phase are :

- jaundice,
- hepatomegaly,
- dark urine,
- light-colored stools,
- etc.

5) What are the symptoms of the convalescent phase ?

- The symptoms of the convalescent phase are :

- jaundice decrease,
- the liver returns to its normal size,
- and the color of the urine and stool returns to normal.

6) What is another name for the posticteric phase ?

- Another name for the posticteric phase is the convalescent phase.

Chapter 10

1) What is hepatomegaly ?

- Hepatomegaly is an abnormal enlargement of the liver due to inflammation of hepatitis.

2) What is steatohepatitis ?

- Steatohepatitis is the inflammation of the liver caused by the accumulation of fat in the liver.

3) What is an acute liver disease ?

- An acute liver disease is a disease that lasts less than 6 months.

4) What are the symptoms of acute hepatitis ?

- The symptoms of acute hepatitis are nausea, vomiting, fever, jaundice, etc.

5) What is a chronic liver disease ?

- A chronic liver disease is a disease that lasts more than 6 months.

6) Can hepatitis lead to brain disease ?

- Yes, hepatitis can lead to brain disease.

7) What is encephalopathy ?

- Encephalopathy is a brain disease.

8) Can hepatitis be diagnosed with their symptoms?

-   No, hepatitis cannot be diagnosed with their symptoms.

9) Why can't hepatitis be diagnosed with their symptoms ?

-   Hepatitis cannot be diagnosed with their symptoms because the symptoms of hepatitis are non-specific.

Chapter 11

1) Should alcohol be avoided in case of hepatitis ?

- Yes, alcohol should be avoided in case of hepatitis.

2) Why should alcohol be avoided in case of hepatitis ?

- Alcohol should be avoided in case of hepatitis because alcohol is toxic to the liver and because it is metabolized by the liver.

3) Why should a patient with hepatitis consume a low-fat diet ?

- A patient with hepatitis should consume a low-fat diet because it reduces the work of the liver.

4) What is enteral ?

- Enteral is what is in relation to the gastrointestinal tract.

5) Give examples the enteral route in the body.

- Examples of the enteral route in the body include :

- the esophagus,
- the stomach,
- the small intestine and

- the large intestine.

6) What is parenteral ?

- Parenteral is any path that is not in relation to the gastrointestinal tract.

7) Give examples of the parenteral route in the body.

- Examples of the parenteral route in the body include :

- intravenous,
- intramuscular,
- subcutaneous,
- intraperitoneal and
- intradermal.

8) What are the functions of the spleen ?

- The spleen filters blood and fights infections.

9) What is splenomegaly ?

- Splenomegaly is the abnormal enlargement of the spleen.

10) What is the function of albumin ?

- Albumin maintains oncotic pressure.

Chapter 12

1) What liver enzymes are important during lab results for hepatitis ?

- ALT and AST.

2) What is ALT ?

- Alanine transaminase.

3) What is AST ?

- Aspartate transaminase.

4) What is the normal ALT value ?

- The normal ALT value is 7 to 56 unites per liter.

5) What is the normal AST value ?

- The normal AST value is 10 to 40 units per liter.

6) What is the normal bilirubin level ?

- The normal bilirubin level is less than 1mg/dl.

7) How is ammonia produced ?

- Ammonia is produced by the breakdown of protein.

8) What is the normal ammonia level ?

- The normal ammonia level is 15 to 45 μ/dl.

9) Is ammonia toxic to the body ?

- Yes, ammonia is toxic to the body.

10) Is a high ammonia level a sign of hepatitis ?

- Yes, a high ammonia level is a sign of hepatitis.

11) How can a high ammonia level in a patient be corrected ?

- A high ammonia level in a patient can be corrected using lactulose.

12) How is the body protected from ammonia ?

- The liver converts ammonia into urea which is then removed through urine.

13) What are the signs and symptoms in a patient with a high ammonia level ?

- The signs and symptoms in a patient with a high ammonia level is confusion, a change in mental state,etc.

Conclusion

Thank you again for purchasing this book. I hope it has helped you in your journey to understanding Hepatitis and how it affects the people around you who suffer from it.

Thank you.

www.ingramcontent.com/pod-product-compliance
Lightning Source LLC
Chambersburg PA
CBHW031510210526
45463CB00008B/3180